I'M DEADLY SERIOUS

I'M DEADLY SERIOUS

Chris Wallace-Crabbe

Oxford Melbourne New York
OXFORD UNIVERSITY PRESS
1988

Oxford University Press, Walton Street, Oxford OX2 6DP
Oxford New York Toronto
Delhi Bombay Calcutta Madras Karachi
Petaling Jaya Singapore Hong Kong Tokyo
Nairobi Dar es Salaam Cape Town
Melbourne Auckland
and associated companies in
Beirut Berlin Ibadan Nicosia

Oxford is a trade mark of Oxford University Press

First published as an
Oxford University Press paperback 1988

British Library Cataloguing in Publication Data
Wallace–Crabbe, Chris
I'm deadly serious.
I. Title
821 PR9619.3.W28
ISBN 0–19–282131–8

Library of Congress Cataloging in Publication Data
Wallace–Crabbe, Chris.
I'm deadly serious.
I. Title. II. Title: I'm deadly serious.
PR9619.3.W2814 1988 821 87–23990
ISBN 0–19–282131–8 (pbk.)

Set by Wyvern Typesetting Ltd.
Printed in Great Britain by
J. W. Arrowsmith Ltd., Bristol

for Marianne

ACKNOWLEDGEMENTS

Some of these poems first appeared in *The Age*, *Age Monthly Review*, *Antipodes*, *The Australian*, *Australian Poetry 1986* (Angus & Robertson), *Meanjin*, *The New Oxford Book of Australian Verse*, *Neither Nuked nor Crucified* (University of Newcastle), *Nimrod*, *Overland*, *Oxford Magazine*, *Phoenix*, *Poet's Tongue* (ABC radio), *Rattling in the Wind* (Omnibus Press), *Southerly*, *The Times Literary Supplement*, and *Verse*. Grateful acknowledgements are hereby made to the editors in question.

CONTENTS

I

II

III

I

THE WELL-DREAMED MAN

And in my dream I woke from the one I was having,
a dream which was nothing more than mere,
began to walk up that S-bend gravel track
which I first pictured some eight years ago,
lay down in thick sweet grass like a boy in a painting
with my straw hat over my face
and began to enter the dream which includes all dreams,
which proves at once particoloured and spacious
with a grainy feel to it like old floorboards
but gets confused with a black-and-white film I saw once
or think I saw but maybe only dreamed of,
so vivid prove its memory traces tonight
while belated cars groan fast along Brunswick Street
less than a block away
with hearts in their mouths
as though they had something big to offer knowledge.

DRY GOODS

It is surely winter;
the sun spills like lemon juice;
I danced last night
having sung under claret the night before
but all this arvo
as a clean wind goes combing wattle blossom
I seem to be a residue
of chipped terracotta
that an archaeologist has found
in a tumbledown tomb.

Don't worry, serious scholar,
it may be that I contain
two dozen ears of wheat
to grow
                    then ripen again
three thousand years late.

Not yellow light but some faint smell
under the north wind's fingernails
whips up
one of my oldest follies,
the firm illusion
that I will live for ever.

Having said even this much
I begin to feel triumphant,
as hairy and assertive
as big banksia heads displayed in a vase,
and could make a cardboard model of paradise
from some old cereal packet,
glowingly in possession of
the gaze, the voice, the nothing.

Don't worry,
it is surely the case
that we shall be reborn
golden as butter on a chunk of new-baked bread . . .
but where? . . .   in what form? . . .
and how can we know that we are?

INTENSIVE CARE

You can always tell the sisters from the doctors.
Passing through glass doors for the umpteenth time
I shuffle it all vaguely together in shock:

Trolleys, people in parks, people parking,
Imprecise reference to supposed registrars,
Gorgeous darkgreen of early autumn foliage,
Towers of midcentury functionalist dreck,
Martian citizens deep in their cargo cult,
Life as usual on a knife-edge or standing prick,
The wards now falling behind for a while.
Making the breakfast, a meeting, getting the washing in
Through a world of participles, not active verbs,
As passive as a bald tyre in a puddle
Or a clue in a puzzle, if it comes to that,
Which it always does. Come again.
                                                            Which it always does.
This is my tram now with the usual number
Hoisted up on the front there, under glass.

I'd like to be as funny as a real hexameter,
Confident as Charles Harpur or Billy Hughes,
Rolling up my sleeves and getting my hands pretty dirty
While knowing that Marx is all just cafe talk—
Where theory sprays a duco on lingering bullshit,
Giving it the gloss required for promotion.

Knowing the faun's arvo for an old dog's breakfast
And seeing the trees in plateglass offices,
I fold the hospital away
                                        again
                                                    somewhere
Rocking its cargo to sleep.

The suffering is real.
Oh yes, the suffering is fair dinkum
In a funny way where nobody gets the joke.

THERE

At the bottom of consciousness there is a clear lake
The waters of which throb ever so lightly
(Like the bodies of lovers after their spasm ends)
Throwing dimpled distortion across the rocky bed,
Greenish round rocks, the size of a grapefruit, say,
And through these cold waters fish are swimming
Seeming quite continuous with their medium
As sexual love flows directly through God.
Here water moves the slubbing barabble of language,
Gust and pith, cacophony, glossolalia,
Gift of the gab and purple rhetoric,
Moaning in rut, scream, snicker, and the rip
That is sheer pain.
                    Yes, these are of language
But not yet *it*. They are the pool,
Its diamonds and yabbies, ripple and scale,
Insatiable glittering . . .
            I'm afraid I don't know what paths
Lead up from the pool to where I think and talk;
By what stony track with landslip and synapse
Distracted everywhere, choked with scrubby thorns
We got to where we are. Conscious.
                                        Aren't we?
Oh hell, we seem to think we understand:
When I ask at the ticket-box they sell me a ticket
But I do not know what the recently dead will ask us
When they walk through the scrub again like sunbeams.

THE STARLIGHT EXPRESS

If you can slip
your fingernail
under reality's near edge
you then begin
to hear those other things:
a child's yawn offstage
crickets tuning up
in bootblack hedges
swash and drag
of immemorial surf
clock all bustle
fridge idling away like a tired hound
but a bird somewhere
enjoys an idea
and thin bright stars
in their savage beauty
whisper secretly
'You must die.
You know that.
You have to die.'
I know it, I riposte
uncomprehending
but please do not let them
burn my body.
A moment of pure silence
grips all nature
like the scepticism
of angels.

STUFF YOUR CLASSICAL HERITAGE

Gull, grevillea, galvo, Gippsland, grit—
just singing out the chorus, bit by bit
will get me some purchase on the primal scene.

What do I say by seeing and then saying
a ragged strip of bark rips itself off
the slender limblike trunk of a manna gum

with faintest crumpling noise?
                                        Call it said,
call it commitment to a twiggy particular
fawn-scumbled slope or terracotta roof.

By naming, I seem to crush the past
like a mattress, hard down in history's
rusty cabin-trunk: stick it in the cellar.

In a way, I preach the destruction of Europe,
that mental Europe which I love so much.
Cancel it. Smother it with ripe new words

or old ones triumphantly misapplied,
every solecism a seal of triumph
as light gilds a scraggy bacon-and-egg plant.

Keep Jehovah in his place with Bathurst burrs
where things are wiry, scrabbled, porous, drooped
for Oedipus romping through the undergrowth

every bit as gaudy as
those three dippingly quick rosellas
or a Violet Crumble wrapper.

OBJECTS, ODOURS

Stale tobacco smoke
gradually gliding
into blood and bone
insidiously
becoming cracked earth

along with
somebody's musty-dusty
awareness of phlox,
nasturtium, wallflower,
a painful pot-pourri

that grips you
clean inside
with its knowledge
of summer grasses, wet spaniel
or kitchen bread . . .

Turn, say,
to the language of wood
in an empty schoolroom
or a webby toolshed's
damp lattice where

hearing
the steady, flowing,
interminable guff
of your grey elders and betters
you find the thick light suddenly

clearing,
dullish objects growing
into a rough concordance: free of your fetters
you catch some crude gist
spot on, gingerly.

THE THING ITSELF

The important thing is to build new sentences,
to give them a smart shape,
to get acquainted with grammar like a new friend.

One rubs down syntax
into a coarse familiarity,
such foreplay as closes down all thought.

Were it not
that the undertaking is too mannered
(as gnostic as a shower of rabbits),

I would like to go right back,
devising a sentence
unlike any such creature in creation;

like nothing on this planet:
a structure full of brackets and cornices,
twigs, pediments, dadoes and haloes and nimbs,

full of nuts, butter and flowers!
sinewy, nerved,
capable of blotches or of waving hair.

That would be a sentence to really show the buggers,
like a cute
new thing

or like a tree
recently invented
by some utterly brilliant committee;

it would glitter, articulate,
strum and diversify.
It would be the thing itself.

RONDO

My muse is not a meson but a gully
Splashing just when the brainpan drowses dully.

My muse is not a gully but a goddess
With all the eucalypti in her bodice.

My muse is not a goddess but a slattern;
I specialize in lack of any pattern.

My muse is not a slattern but a glory;
The light she lent was merely transitory.

My muse is not a glory but a critic
Whose role, I fear, is glum and parasitic.

My muse is not a critic but the phoenix.
I catch his burning teardrops in a Kleenex.

My muse is not the phoenix but a meson
Beyond the pale of measurement and reason.

MADRAS

(*for George Russell*)

The palms had all those things to say,
the greenish, purplish, dusty palms
clattering like venetian blinds;

quiet the evening, busy day
as dhotis flapped along our road:
dust and bananas I recall.

The palms had much to say,
they, and Ramani's bamboo flute
locked into centuries of calm

or else of more than calm, or less:
something other, anyway,
that high-coloured city had to say.

Not just crammed folk, the English store,
the squirrels or the dairy shops,
beyond five trishaws something else

might have been there
or definitely existed there
as calm as coriander, say.

We loved the place, but could not say
what silk or saffron, what thin tree
distilled all genius of the town,

it so much haunts the mind . . .

A GLIMPSE OF SHERE KHAN

So the jungle
was pard and barred,
reflexive with shadows
patterning our path
as the coir saddle rocked slowly.
We were all eyes.

Then Bashir, mahout,
straddling behind the beast's ears
gripped my wrist hard.
'Tiger,' his teeth hissed
as our elephant began
a long, strong bout
of shuddering through and through,
head on to where
stripery had been and gone.

Around, back and up right,
canting to cut him off
in some douce clearing
and—
        HIST!—
                so we did:
all aglow now, his yellow
astripe, approaching, grand,
he had us heart in mouth.
We stiffened. He stood,
shade on his hide,
the whole dry glade
gone tense. It seemed an age
then, flick, he was lost,
filtered through thin jungle
slotted in space-time.

CITY

(for Bella and Boris)

Outside my double glazing
erratic snow
came dithering upward
like blown flower petals

but on the bridge's back
it turned aggressor
and pelted a hard confetti
into my strange face.

The river was black as barges.
All the ziggurats of light
blurred painfully lyrical
as memory has done.

Just about any unknown street
is crammed with escaping lessons;
every fact we have grasped
we forget and die.

I'm afraid I am a tourist
visiting my life, where
all the postcards look downright lovely
and I am still afraid.

From a little park's flat chest
to the bridge's back
this pitiless snow accrued
like compound interest,

a vast exciting monotone,
imperiously white on white,
prolific semen of oblivion.
But night blew in:

on the river's rich black skin
jewels have been thrown to comfort me,
diamonds that tremble for love
and gold, a pledge of hope.

VENUSBERG

The fluent grassheads rumple
like a silken summer bed,

elegant walls of blackberry
are woven in arabesques

and someone has daubed in a bluish backdrop
of buzzing western city.

Yellow signifiers of wattle bloom
have come and gone, every bit as transient

as champagne or last year's cherries
but ambience lingers with a human force.

The ground has thews and haunches. The sky
has an eye. This little savannah

can pulse like very Mozart, holding meaning
in the kiss of the palm of its hand.

THERE IS ANOTHER WORLD BUT IT IS IN THIS ONE

When the unspeakable has taken place
or even almost
landscape cannot help,
music is up to no good

and food cannot pick you up,
balanced on brink of nausea
with bellygripping pangs,
shitting yellow like a baby.

Books and slog do their level best
but cannot help for long
as the compass arrows home
to that knob of horror.

Try, need, to grapple sleep
and you wake on your sodden pillow
tacky as plum jam,
then shiver your arse off.

The answering bay of calm
is probably religious
but you lost your way to all that salt coherence
two centuries ago.

A bloke walks tentatively
in my direction carrying
the yellowed skeleton of a bulldog
on a polished plank

and not a bloody thing
can, yet, now, quite stop me feeling
we are connected to a coursing stream or tide
incomparably larger than ourselves.

It is pure thought of the water
forky with wreathing rivulets
calms that sickness under my heart
until

I tick-tock once more, rue, recall
it was one of those ropy spangles
threw up like a dice
the fate-blow, and all seems again

like Darwin's tangled bank
whereon the lovely millefleurs blow
round killer birds or insect wars
malign as melanomas.

What is the crossing or clew between
that which means god
(that which God means, comedian)
and this cheeknipping brown souwesterly

puffing between trading banks'
renaissance-resurrection porticoes
flaunting their red granite colonnades
and rivery marble cheek?

GENIUS LOCI

When I can't sleep and prove
a pain in the neck to myself
I will sneak downstairs, dress up warmly
and squeeze into whelming darkness
or piccaninny daylight, where
I may just glimpse at a corner
one of the Jika Jika slipping away
lapped in a possumskin rug.

I will hurry like steam to the corner,
ever so much wanting to say,
'Hey, wait. I have so much that I . . .'
But there will only be
broad street, creamy houses, dew
and a silence of black shrubs.

Maybe if I got up
a little more smartly next time,
got out on the road quick,
I could sneak up closer
on that dark tribesman in his furry cloak
and ask him . . .
                                        oh, something really deep:

something off the planet.

HINGE

As petite clouds
go fluttering prettily
over that highrise wedge of sky
Koori people
with plastic bags
crammed—perhaps—full of mystery

sit hunched or slummocking
on hot benches
at the knees of their health service.
One of them
slopes over to cadge
a cigarette if I have one,

but a dollar would do,
or history reversed.
There is geological spoor
under these black footpaths
and the past
keeps heading west.

While March sun
on these benches
burns a going century clean out
of our bones
the Second Coming
snoozes under chalk-blank newsprint.

Ah, dingy Gertrude Street,
the terrors of the earth are never quiet
and lost blue tribes
are dancing like the sound of crabs in a pot
on our thin coffin roof.

Forty cents will do
for a morning paper
from which to find
whether a bombing raid on Benghazi
proves much the same sort of thing
as it did in my boyhood.
No.

THERMODYNAMICS

'It was not power that you lacked, but wishes.'
JARRELL

I am shivering all over
Because we drove to Cambridge
Where those haranguers came from
Who did us all so much harm.
I curse the whole pack,
Those narrowgutted turds
Or bookburning moralists,
Spitting them out of my mouth
For burning so hot and cold
By academic rote.
Put them in a footnote—
All crablice from the damp
Groin of history—
And let us chip a few
Lines to biology,
Insatiable virago
Who keeps our chromosomes
In her suspender top,
Laughing the sparrows to earth,
Sunny to see them drop.

I'd like to believe in something
Much larger than myself
But all the designs are crook
Or disproportionate
Or just more bloody art:
I've been using this hand for years
Yet can't describe how it looks,
Nor even the shape of a tree
So you could see it at all.

I am shivering down my bones,
My very thighs are chilled . . .
How time bends and deflects.
I was thinking of dumb B
Just the other morning
But could not even recall
Whether J was his mother or wife.

Biology, crude mistress,
Had conflated those two:
Ah, how she has been
A cunning taskmaster
(With weird incunabula),
Having had the hide
To allow poor penis no brain,
Strong on phylogeny,
Providing for migraine,
Piles, cancer, dystrophy
And galloping pandemic.
We hutch her in our garden
To mulch, to prune and to graft
But she kills us all still
Hanging our flayed skins
On the family upas-tree.

My marrow thus deepfrozen
I tiptoe on hot bricks
To the crematorium doorstep
Perceptibly stiffening up
At every ball-and-socket
And learning presbyopia
As a text in history,
Still roaring the old bush songs
But perfectly out of key.

The time that we need we burn,
So I seem to be warming up
Coming back to the rosily
Familiar truth of objects
Seized by vertigo;
I know how shadow striates,
Right now, the winter lawn
Or this colour of pale skinshine
With the veriest ghost of a flush
As the cheekbone curves away,
But where is soul located?
Or, if you like, who is
The unknowable, hence religious,
Projectionist of self?

Or, if it comes to that
Might spirit find queer lodging

In antimatter galaxies
That we can never detect,
Such being how knowledge is
Or that it (is it?) is,
For I live like poor us all
In terror-watching circles,
Fingers returning to life
By way of pins and needles:
Both rage and thermostat
By an odd conjunction spring
From the hypothalamus.

The time that we need we burn:
How much of my life have I
Frittered away in wishing
For the future to hurry up?
Accept.
                    The important thing
About intellectual systems
Is to use them randomly,
Playing the ball where it bounces
And making the most of the pitch;
Because I feel sly today
I can look out this window and say
All I see there furls
A civil wilderness.

Leaves can be musical notes,
Which, boozy Brennan said,
Are Epicurus's gods,
In some interval of worlds
Following their own delight.
Small birds tune strings.

With passage of simple time
Plus application of
Some cabernet shiraz
I appear to be thawing out.
A toast to the vignerons,
Whatever it is they may buy,
Who have known the cyclic vein
Of that image in Gerona
Where the fanning vine springs
From Christ's cut side

Then arches over his head
So he can reach up
Squeezing a bunch of grapes
Into the goblet he grips
With his other vintage hand.
In vino tranquillitas
Or near enough. So *slainthe.*

Hoyle reckons the creator
Was limited in his power,
That our carbon-based being
Was erected on a harsh
Necessity or doom,
Our dailiness representing
The best of a bad job,
The earth but a minor skirmish.
We are pissing against the wind
But with histamines in our veins.

It was pearl Oxford morning,
Sky like a muck of limestone
But not that appropriate black
Of New College's chewed walls
When the phone burst from Melbourne
To say that Dad was dead.
Gone, then, the world that Wal built,
His concrete-footed fortress
Jutting up in soft bush
Against rill, yellowbox, blackberry
And saffron of winter wattles
Bursting in bright puffs. Adventure
Came down at last to this.
Less and less theatres of mind
Will now rehearse his tales
Of Gwalior, Lebanon
(Already blown to pieces),
Akhtur and Dicky Mountbatten
(Already blown to pieces):
Orient as narrative.
Less and less of his chums remain,
As old as the twentieth century,
With its chevrons, kukris, maps
And the red stain washed out.

If I am shivering now
It is for the deaths of us all:
Such fire as I think of, the furnace
Which turned my father to ash.

Spirit finds queer lodging.
I am in the front line now.

II

'Comedy is everywhere, in each and every one of us; it goes with us like a shadow, it is, even in our misfortune, lying in wait for us like a precipice.'

KUNDERA

SONNETS TO THE LEFT

I

What do I get from progress? I rejoice
In antibiotics, the dental drill, clean drains:
Perhaps a little bit in aeroplanes.
Deutschegrammophon and His Master's Voice
Pump a magical suasion through the air
So that we all have orchestras to hand
Or history on deck like contraband,
Bartok and Bach concurrent everywhere.

Slavery's dead. An evening chill now falls
Rather more golden; autumn shadows flit.
I wander round these white functional walls
Trying to find a frame in which to fit
Large things that progress bundles out of sight:
Grief, awe, terror, transcendent light.

II

(i.m. *Judah Waten*)

As we grind into the worst decade for two
Or three or four, I swing around and see
Your bulky, suited, Russian form push through
This or that minor bookish jamboree
Smiling, and think of old hostility,
Those years when I watched you hard for Stalinism's
Cloven hoof and you (I'm sure) marked me
As bourgeois formalist. Time burns the isms.

Now Toorak and Balmain contrive to read
Marx as ur-text, perfect aesthetes delight
In cushioning off him too. The game is bent
Till we've become old colleagues with a need
For shoring words against the tide of night,
Praying the slow bitch History might relent.

III

There used to be a time, as I recall,
When 'Left' involved the proletariat;
I like to think that might be where it's at
But fear it's grown up theoretical—
Or else gone troppo and turned Libyan.
For years it strove with muscular intent
To keep the ALP from government,
But we've now endorsed that gross amphibian.

Soft acker vikings cross a phantom sea
Hoping to put old systems to the axe
And loading up their vocab, trend by trend;
Joying in papery perversity
The deconstructors lie down on their backs
And let our Marxists rape them, in the end.

IV

The writer depersonalizes his dreamwork,
A single being in search of voices,
But change is learned from the outside world
Preoccupied with drink and kissing.
Unhappiness is the funniest thing,
It is a play about the ego
Needing to choose a lifelong project
Through tea and comfortable advice.

How can anyone be called guilty
From the perspective of the trapped
And struggling fly? Perhaps the itsy-
Bitsy is proof against despair,
Knowing between the fork and the knife
All our beyond is in this life.

V

We look edgewise
Through this our dishy galaxy
And see thick jewellery
Spattered across an empty sky
But our damp cries
Balloon at last through history
Which does not leave for personal foolery
So much as a footnote or sigh.

What'll we do with the mystical
Now that red buttons think beneath our thumbs
And secular hope is built upon a rock?
Turn to the fantastical:
In furious rubber last of all there comes
The hooded cyclist with the duck-beak cock.

VI

Look, mate, you roll new brands of foreign cringe
With which to snow the members of your branch.
You are a proper old conceptual stinge
And where you gesture at an avalanche
We only see the dreck of magazines,
Half-baked ideas, half-read, near understood;
You reckon you have proved you're full of beans
By squeaking, 'Foucault', through the sacred wood.

Cling to your books and views of bin-end booze;
That way you won't get dogshit on your shoes
Tailing along to popular festivals.
Look,
        something real is happening out there
And when you finally notice, then, I fear
The feminists will have you by the balls.

VII

Tell this to Yale and Paris, tell Pan Am,
Bloody imperialism never ends;
However much Big Brother dubs us friends
He'll roll us up and dump us off the tram
At the drop of a hat. Ah, *dump* . . . yes, dump:
That is the linchpin of imperial trade
When the time comes to slip the masquerade
And knife your friends under the doom flag, SLUMP.

Australia first, and last, and in between
Australia once again, is playing safe,
Even for us, the rigorous aesthetes.
High art begins from kids' play in the streets
And local heroes earn the long last laugh:
Ocker, metaphysical, obscene.

VIII

Through a green, social afternoon
As to and fro the emptied bottles roll
This question rises, bleary, out of tune:
Can a fuckwit have an immortal soul?
When shall the vulgar court the lyrical
And bullshit fertilize the laurel crown?
When will cute images delight us all
With ratbag stylists going on the town?

Only when stories happen on a seam
Whose gist gets memorized from north to south,
Swimming through mythology like a dream:
The dingo with a baby in its mouth.
Fuckwit and smartarse, trendocrats and folk,
Nothing unites them like a dingo joke.

IX

De Tocqueville said (he really didn't say)
That in the art of democratic nations
The brittle courtly forms will shred away
And be turfed out by more offensive fashions.
O.K.
      Sing of the footy, sun, red wine,
Green surf that batters on this ridgy coast
And then, as redolent as turpentine,
The spirit of the place: our compound ghost.

I like this rhyme, so how about a toast
To that bold Rhadamanthus, tough Jack Lang,
Our patriot whatever (oops!) the cost,
Whose name, returning like a boomerang
Led to this rough platoon of sonnetry
Affirming Land Rights and democracy.

X

Sweet as a clarinet the dawn returns;
I hone my slang aubade, damn sure that our
Priorities are not the Comintern's.
Land Rights lie deeper in the soil than power,
Being the native debt, profound as trees
Turning metallic leafblades to the sun
In morning's natural order. Pieties
Deem that this act of ritual be done.

News twangs a bleeding knell for every age
Or else a snotty-grey decline, unless
We've laid out clear proposals down the page
To crank ourselves out of a fiscal mess,
But if ideals have something left in hand
Tribal sites will peacock a beige land.

THE HISTORY OF THE WORLD

Froth, nodules, granular dynasties,
a glacier dragging
its non-existent heels
artistically through rock.

SPORTING THE PLAID

Renowned as Black Geordie
my Arbroath grandfather
sprang fully armed from the same ten years

which threw up Hardy, James and Furphy:
no wonder I dote on their humour,
that scathing irony.

A burly snob, he hadn't a clue
what to do with his sepoys
till the Raj wiped out his commission

leaving him a helluva lot
of other irons in his furnace,
his friends reputedly Brunner & Mond,

César Franck, Anatole France
and Milan's grandiloquent 'Mr Green'
whose arias he continued to adore.

Blown to Australia, the old buck founded
a blague of Caledonian societies,
tatty diamond mines and a second family

with a blonde Highland schoolie from Cork,
spieling his impossible tales
of clan exploits

along the brown Jumna, on blue high seas,
in a Boxer Rising where
Crabbe's roughriders exerted

their mongoloid talents
ensuring the flood of opium
for a smoky god and fleshpink empire.

I suspect you of being a shit
but in stiff, perfect photographs
magnificent beyond belief

on the bridge between dandy and warlord.
The whisky fed your moustaches,
your children adoring you,

scared stiff, bloodthirsty, tribal.
When you were half-seas-over, roaring under the stars,
'I could break that cabhorse's neck

with one blow of the heel o' my hand',
you were all huff and puff, a bolt of plaid
woven from dropped names.

THE MIRROR STAGE

I met this little girl
One lunchtime, years ago;
Well, she was a widow then
But she had been a little girl.
We used to call her Missy.

And she suddenly spilled the beans—
The widow, that is, in our Staff Club—
That back when I was a boy
They were told not to tell me
My father was killed in Burma.

He was missing four months or so,
Painfully trudging through jungle,
Shot at, magicked across deep rivers.
When the Japs bombed his last jeep
They blew up tins of pineapple.

My father had not been dead,
Just on a long, slow detour.
He savoured three years more in Asia,
The vine of his life, it flowered,
Never to be tasted again.

Years of mild swashbuckling, miles
Of Indian wanderlust:
A war like technicolour.
Now he is burned, and mother
Is burned, and we've buried my aunt.

My limbs are being cut off
To make me hobble faster.
I am waltzing on their graves
Like a sunstruck hatter,
Like Indian ink.

SEX IN THE HEAD

Now that the billboards have been pulled down
along pleasure's concrete motorway
    our dark philosophers
are filling the future with words,
none of them signifying a hoot or a twig
    though turtledoves rasp mildly
saying, it's hard to be good, and
it's got to be hard to be good,
all those quips apparently meaning
    only the queer machine
talking through us, who are its occasions.
    It comes as it likes
as a thief in the night riding his palomino
swish in the black pelt of love.
That I have been rapt, ecstatic even,
through satin evenings on this planet
is something the English language
    has taught me to say.
I do hope it has been happy too:
its verbs and conjunctions quivering away
    through picnics of *joie de vivre*
or laid out flat in some syntactical orgy.

When I say goodbye at last
to this bluegreen ball in space
I trust it will remember me
    as a brownish culture-site
where it put on the odd extravagant production
of *Cosi fan tutte* or *As You Like It*:
as a small quiver full of chromosomes
by then returned to the Public Reading Room,
    the dome of which
has ways of refining our sadnesses
and bottling them under the label, FOREPLAY.

BINARY

Why does a cauliflower so much resemble a brain? All those pale curved protruberances and hillocks tease the mind into activity . . . at this point I randomly remember the complicated architecture of a particular gothic dream. But brains and cauliflowers, ah yes, is this the same kind of parallelism as that which holds between pine cones and pagodas? There we go again, seeking order or duplicity in the stubborn universe. We ask ourselves whether the resemblance between a rose and a cabbage is like that between a clipped hedge and a high tin loaf, coming up with no answer at all. The brain and the cauliflower continue to rise up on their cortices, bubbly fruits that they are. Moon goes around earth goes around sun, et cetera, analogues active everywhere. Echo redeems Narcissus, shadow is touching reflection. We ask ourselves what it all signifies. Somewhere, in shadow, aged sages debate such questions on a lawn all day, over their wine and bananas.

DUSK

The last gone
tennis player's
jumper sags
down the sodden
white wall;
rambunctious
children are
being packed off
to bed,
hopefully.
That's how
the beautiful she-oak
finds herself alone
by the sea
with her
little grenades
and her shocks
of string
music.
And one yellow
ball snoozes
in a gravel puddle.

SONG

Grumble away, sweet ocean,
Your surf comes rolling fast
With such a lyric motion
Your powers go to waste
But the bluegum-shaggy mountain
Ignores your breaking;
Rosellas here are waking
And slowly, very slowly,
I am waking.

Now the bright sun rolls higher,
The ocean falls to rest,
Ideas have lost their fire
And reverie seems best,
But the pulse of coastal music
Grows oddly firmer;
Thin creeks and gumtips murmur
And, pushing through the ti-tree,
I, too, murmur.

IN THE OTWAYS

That nannygoat
contrives to sleep
with her white head
quite vertical

luxuriously
rubbing her chin
on the rough bark
of a pine tree

lullabied
all the green while
by one melodious
native thrush

going all through
its repertoire
to a drunken scent
of bluegums.

THE MIDDLE YEARS

In, past or beyond
the topaz-tinted windowsquare
beside the station

seen through cyclone fencing
a big fattish man
in creamy singlet

pulls from his wardrobe
a flat plastic bag
holding grey trousers,

puts them on and douses the light.
Beyond that barn-red
galvo roof with one grey sheet

stormclouds mass
heavily in hanks
under a threequarter moon.

The house is dark green,
forestgreen perhaps,
and putty flecks the windowframes.

Nothing else happens
and my train is roaring in;
I hear it now.

Red and green mounted lamps
mark the two platform ends
with opposite meanings.

THE BURNING BUSH

Almost opposite, a rhus blushes outrageously.
In the midst of all these houses it is like a flame,
if flames were ever so richly, darkly red,
if flames were ever held to bedeck a garden.
Its colour is too much, it deepens the autumn
and sky withdraws in pastel scarves behind it
making those gestures that prepare for sunset
while a hidden sun puts fringes
around a cloud shaped like Java, a pewter cloud.
Though tarted up by autumn and flecked with puddles
our street is finally rigid as politics,
a good place to settle down.

My mother, were she still alive,
would have delighted in these alien colorations,
such daubs of japonaiserie.
And now, like punctuation,
a little old Sicilian woman
comes beating back up the street
with her plastic bag from the deli
and the bright Virgilian leaves blow round her feet.

They dub this the first day of winter
and I have slipped into my autumn
so I'm allowed to do anything I like,
as long as I'm up to it. The leaves hanging down
from these gutter elms are brilliantly yellow now.

STARDUST

To a smell of water-vapour and wood fires
Walking by night, my breath allegro
To feel it's all not worth a cracker . . .
But how could the universe have meaning?
Would the stars be patterned differently?
The seasons vanish, or come on faster?
Would there be an End?
Perhaps we wouldn't require any sleep;
Maybe we'd no longer have to shit;
Or one radiant mathematics
Would show up trimly in everything.

Everything is just as bright
As the hollows are Indian ink
While my bones go wandering chockfull
Of a crushed silver.
These paddocks have all been marked out
With expressive diagrams
As beyond the highway ribbon
Big waves crumple and bang.

But when you ferret after the meaning
Which a universe could be hoped to have
(Oh dear, yes, one has a cold,
Or has an exam the following day,
Or on occasion has an erection,
Or they have a holiday shack up the bush)
You stick at a sort of spatial problem:
Meaning is only a bundle of signs
That parallel and light the real,
But would they then be *in* the real?

Pan has left the odd footprint
On somebody's wet lawn
And his hot metallic stars
Are doing the rounds of my arteries:
I can even feel the moon
Down in my quicksilver groin,
The drypoint shadows
Falling across my brain.

Then signs are doublewise at once,
Being inside and outside what they picture;
If not, they're simply beyond our ken
Like God's hand moving among the stars.
We must find a little enclosure for meaning:
It needs living room, like a dog or a student,
And won't be satisfied with my brainpan.
I hope we can find a cosy gap
To bed it down in after all.

In those juvescent nights of starshine
When I knew what the future signified
A full moon could
Shake me with stony horror
But now it hauls me into
Pure aesthetic compliance
As a pinecone rears over its shadow
On the concrete pathway.
Such fallen silver as this
Leaches down through cracks in the earth
To take the place of marrow in ancestral bones.

III

for Georgia

'The emu-section of the dream being thus partly fulfilled, Bill clutched at a release in any form.'

FURPHY, *Such is Life*

GOD

That is the world down there.
It appears that I made it
but that was way back,
donkey's years ago, children,
when I spoke like a solar lion
beguiling physics out of chaos.

I spun my brilliant ball in air.
Such thought was new to me
though I had not guessed at my lack
in the old indigo days,
children, before you fell—
to use a technical verb.

It is full of beautiful flair,
a jewel and a garden at once,
bluish-green with the track
of silver engraving its veins . . .
Shit, but it's lovely
and no end of trouble at all.

Children, it once was bare
of all salacious language,
of goats and bladderwrack,
of banksia trees and wrens.
I endeavoured to bring it up rich.
I reckon it's my museum.

I gave a big party
and the name of the party
kept slipping clean away
from my wooden tongue
but I reckon it was
called history.

Some honoured guests
took off their names
or left them impaled like scarecrow rags
on my staggy front hedge.
I thought of it as being
a party for my son.

WHETHER THERE IS TERRORISM IN HEAVEN

When Satan said to God
beside the lichened wall of Paradise,
'I always felt that death
was the greatest of your little jokes',
the Lord picked his nose,
I suppose,
seeing over Satan's sooty shoulder
not a lazy lizard in the sun
or parrot on a twig
but the new, burned peasants
taking cancers to their bosoms
like a bride
and the poor plough twisted
in a fruitless furrow
not far from a bend in the highway.

DOWN UNDER

Someone goes walking over my grave
And I can't quite tell who.
They have filled my eyes with sandy loam,
A bloody thing to do,

But I suppose I wore them out
On scenes of little cheer—
Chicane, the jobless, grasping greed—
Before I fell down here.

The sons of god are lurking
Behind that pittosporum hedge,
Their haloes and their sandals
Heaped on a window-ledge.

Some snooper has come to a shuffling halt,
Treading the bone-dry fescue
Above my belly, my cock, my feet
As though he considered rescue

Or she did. Doesn't matter much;
My lips are long unkissed
And safe down here I think of myself
As a flat geologist.

Tectonic plates slide silently
Between here and Peru;
That scungy pub just over the road
Is called The Drop of Dew.

Whoever is walking over my grave
Through Bathurst burrs and hay,
Leave me. Turn your key in the dash
And quickly drive away.

THE LORELEI

Some have seen her by the highway
  From Bordertown to Nhill;
As they leave a purple sunset behind,
  There she waits, quite still.

She's been spotted down in Gippsland
  Or along the Great Divide
In the shadow of a redgum
  Hitching for a ride.

Straw-blonde, in a windcheater
  With frayed and faded jeans.
Some think she's in her early thirties,
  Some in her late teens.

You are driving towards your future
  Doing eighty down some road
When there she waits on the grassy verge,
  As motionless as a toad;

You ask her if she wants a lift;
  She smiles and sits in the back.
She's only got a handbag
  And a smallish haversack.

You exchange a few small pleasantries
  Then conversation stops
And all around the narrow highway
  Comforting darkness drops.

Another dozen miles and you feel
  A pricking of your hair.
You look in the rear-vision mirror.
  There's nobody there.

There's nothing at all on your back seat.
  By now you're spooked as hell:
We suggest you cruise on carefully
  And stop at the next motel.

WALTER BENJAMIN

(d. *Port Bou, 1940*)

How years called back those nursemaids
who rolled him to the city,
his maps and ineptitudes;
all those lived streets and kiosks
wove a sexy solitude
through whose dirt Berlin became
Naples, Marseilles and Asja.

Fattish, slow, solitary
and so randy for bookshops
that he lost his way to them
this laureate of aura
found the little hunchback's gift
left there by his bassinet,
bright, haunting and boobytrapped.

In wooden Moscow he slept
at a ramshackle hotel
where almost all the rooms were
taken then by Tibetan
lamas who didn't shut doors.
But through a beerhall door
comes the harlot, Memory.

A glimmering shone through panes
in which a (or *the*) woman
floated with nutbrown eyelids,
precursor of that Leda
who received the swan between
her parted thighs in topaz:
a call wakens an echo.

Our capacity for ex-
perience is decreasing,
he complained, quite ravenous
for objects in displacement
and a free, massive sadness:
perhaps in this century
we cannot *own* our feelings.

Beyond, he must always go
beyond, even pressing past
aura's peregrine taproot,
skirts of sepia, finding
in Hill's Newhaven fishwife
'a seductive modesty'
and a 'how did that mouth kiss?'

If all the decisive blows
are struck lefthanded, God knows
who struck the blow at Spain's edge
by the fatal Middle-Sea.
Your malign dwarf trapped you there,
claiming complete disorder
on this side of the border.

Alongside our bumping panes
rose the grass-furred, sheepgrey hills
by nowhere, or the trainline.
We drank the grimy glamour
of a new where.
                                        Stopped.
                                                            I mouthed
croissants, coffee, displacement;
this was breakfast at Port Bou.

JULIE

Somehow, I don't know how it is,
I want to keep escaping
From those little green guardian eyes
The word-processor displays.

I keep on gazing out the window
Past cars and other windows,
Waiting for the prince to ride up
And bear me squealing away.

It's not that I'm not independent—
Oh god, I am so stubborn—
But I look straight through the typing pool
And wait for the prince on his charger.

You know, dreams are more terrible
Than any kind of late night movie:
I dreamed he rode up as a Mongol khan
And thrust his lance into my side

And then, as I lay on the stony turf,
He took pity on me, bent over
And filled my mouth with roses
To bring me round.

Mum reckons I'm a regular dill
When I forget my watch again—
But a watch only means time.
I just love getting flowers.

Neil brought me an orchid that night;
We had sex in the back of his car
Afterwards. He was rather drunk
And it wasn't graceful at all.

Suzie is different. She cares,
I suppose you could say we're best friends,
But after the office Christmas party
She rested her hand on my bottom

Just for a moment. That was all,
There's nothing queer about lazy Suzie
But a wonderful electricity
Ran all through my body,

Not at all like sex, but like the prince
All the same, putting his gloved hand
In a pot of gold at the rainbow's end
And pulling out sapphires.

Nothing has happened since, not quite;
But I think I'm happiest when
Suzie and I drive down to Red Bluff
And swim straight out to sea,

Dive, splash, float on our backs
Or just plain horse about.
She has the most fantastic bathers,
Navyblue Speedo, smooth as skin,

While I feel at home in bikinis.
Tastes are all different, funny.
A summer beach is the life for me,
I feel happy as going to Mass . . .

You know, I think that nature, sort of,
Bare skin, sun on the sand,
These are the things that are most religious.
Perhaps I'm just a cavegirl,

Which is what Dad reckons about me
When the dear old thing isn't doing his loll
And snapping, 'Keep your mind on the job
If you ever want to succeed.'

Wherever 'ever' is, I reckon
That I'm probably there already, with bells on.
Tonight I'll go to a gig with Suzie
But on Saturday night she has Dave.

GASTROLITHS

The she-fox with her belly full of stones
limps around the trodden hill
where she buried white bones;

she avoids the barbs of gorse
slowly by starlight.
Where is the moon? The horse?

Pebbles of quartz, flint, felspar, malachite
rattle in her gut.
Is she Carmen with a castanet?

Maybe. Ask her.
Rising on her hind paws
she grimly begins to dance the bergamasque

for every fox can tread a measure
or must, in the stilly chill of night. Look,
on the other slope of pain there rises pleasure

albeit slowly. And then
over the starstruck gorse
yellow summer will bend again.

THE ORIGIN OF DRAGONS

Some of them envied the stag his crooked antlers
and some that lithe subtlety of the snake,
some the fire that spewed out of volcanoes,
or Leapy Diver the butting penis,
some that scaly log the crocodile,
others had found huge dry bones under clay,
others a dream to show to Mr Jung
out of the common pool,
some felt that this was how Eve had fallen,
some told good yarns
and others gapingly had to aggrandize Michael & George,
those tinned swordsmen
as armorial killers.

But all of them spun the good old stories,
tales of the gorgeous rainbow snake,
the feathered serpent,
the drugging blood of Fafnir slain
so that a hero could overhear
prophetic speech of birds up to no good
and plunge on to his doom,
or tales of how the dragon's pale soft belly
rested sweating on a heap of gems:
a bloody uncomfortable bed.

THE SIXTH MAN

(a tale of the Cold War)

. . . seventy-eight, seventy-nine, eighty,

like some ink drawing
he plods to the centre circle of a Moscow park,
thick boots crunching ice crystals.
He has fallen a long way from School House

yet that old imp of fun
bubbles up at times from wherever old things go;
last week he graffitoed BURGESS LOVES MACLEAN
in a marbled restaurant loo.

He has been wedged here five years now
in thickets of cyrillic characters;
if all those vodka bottles were laid end to end
Intourist could run trains full of Cubans on them.

Not young now, he shivers over the case of CURRY
and the fiasco surrounding DAGWOOD
in all this whiteness, but was he, himself, and how
a double-agent for whom?

Tracks in the snow his answer;
babushkas passing by under hats
like fur igloos. And this year's cough
growls deeper in his chest.

> Consciousness is irreversible, I thought
> in a hiss of sibilants,
> neglecting through my thinky joy
> all the bleak evidence.
>
> Trouble is, you hear no Purcell over here
> just verst on verst of bloody Tchaikovsky
> like being in Hell—or practice for.

He'd love to sip a pint in a decent pub

who never will, though down by wineblue Black sea . . . ah!
These dreadful birches
are skeletons of something on earth once known
when knowledge existed—silk stockings—the sensual.

They say old Anthony really likes Poussin,
that's Cambridge for you . . .
                                        And who the hell
gave somebody once the cool idea
to crush Mark Dymock with the honey trap?

Prague pounced,
Mark fell, became one of *theirs*,
got promised the sun and stars and crescent moon,
then took a dive under that train. Too soon, too soon.

He can hear Mrs D howling on his shoulder
afterwards: 'You at least were Mark's friend.'
Where have all the birds gone?
Drifted away like Christians or like hope.

Translating daily at the Institute
is a piece of bloody cake.
He plods to the frozen fountain, gone all tense
with wondering if anyone might turn up

waits, clapping stiff mittens, turns clumsy
homeward along a swept path.
Nothing. A square Slav stomping by
then leers at him, breathes, 'Good afternoon, Carstairs,'

in curling puffs of steam:
the accent is Home Counties polo club.
Even a hardened old performer like Carstairs
halts in his tracks, poleaxed.

> *He saw the school close, sunny and warm . . .*
> My head spins like a leaf, out of season,
> my brain's two hemispheres
> tied in knots by their cold war.
> I'll struggle back to sense.
> To be grown-up and gloomy is something

even in Holy Russia, chum.
Up your bum, Guy used to shout,
but fancy being a lad and have
Dad peg out suddenly while screwing Mum.
The traumatized are commonly harum-scarum,
it was Guy gave buggery a bad name,
thieving among the fallen;
joke was, he abominated Russia
but then, I'm afraid, the powerful fairy
who grants us all one wish in our lives
never tells which one was granted.

They clomp forward, flank by flank, like clouds
taking each other's meaty measure
(andante generoso)
for a while silent, that being their pleasure

and then this weirdo blurts out,
'Netta asked me
would I send you her love.'
Netta! The huge streets gape like rigid steppe

and Babushka Anadyomene
is halted in her tracks like any tank
on every side. Who is this wise guy,
where did he spring up from?

A smasher of a party at Sacha's dasha
only last month; we bottled jam
along with this dishy ballet-dancer
from Kiev, her breath Grand Marnier or yesteryear.

Forget that dose of pastoral for
who is this guy,
this fatty bundle of fur and stormgrey wool
with a stockbroker-Surrey accent? . . .

In that far unimaginable world
I was worldlier than Goethe, funnier than Stalin
or Rumpelstiltskin, and when Mata Hari opened up
the honey trap
I hopped inside, stole all the furniture;

I was the Brothers Grimm or Sherlock Holmes
but this infinite cold petrifies a giant city
or life itself, even in autumn.
Then, I seem to have done it all for peace.

Netta—brainstorm:
for cagey years, for my own good
I have dubbed her Mrs D.,
even to my deepest nut of self.

Well, then, Mark fell to his death
and CURRY was blown
when Kim tipped a wink mighty quick.
I've never yet seen a man die,

it's funny,
that is to say not at all.
I wish I could shake this big knowall
bastard clean off my back.

To a bird's eye view
they pace out equability
but his brain leaps like a netted eel,
an ox leans on his tongue.

Under squeeze from Slivocek and the Czech-check boys,
fingertapping Mark Dymock, poor bastard,
had put a bug on the meeting with
some of the Septic top brass

and got away with it;
that's what we all think:
MI5 was already hotfoot after
wobbles in the poor beggar's fancy footwork.

But when Carstairs drove down to Hastings
for sand-in-my-shoes with blonde Netta
his guts were twisted like rope, remembering
Mark's very favourite quote:

'All instructors are just the same.
They tell you to cut off so much fuse.
We double it to be safe,
that's why we are still alive.'

No double means trouble,
everything does,
birch trees like rickety bones
and the newspapers under glass.

One, two, three, four, five,
Carstairs steps it out faster now
by geometries of banked-up snow.
The weirdo lets him go,

for now. Do the old gang
keep their tabs on Netta?
Pack of ruthless bastards.
And what does she do these days?

They had once been basking, flank by flank,
on a beach outside Bordeaux,
he still marvelling at the power
with which she could swim and dive,

when she said, out of the hazy blue,
'And the future, James,
what do you see in the future?'
'I don't know at all,' he had lied.

But the past was another mess and matter
where Kim destroyed Volkov in wonderful Istanbul,
where Satan went to war with Lucifer
and CURRY was dead as mutton,

another lost soul, being one
in a haggard vanguard of self-doomed chaps.
'We only progressed by means of windfalls
which threw the stuff in our laps.'

Nobody; Someone: what do you do
for identity as the sky shuts down,
not to mention mind turning to snow
as the chill past falls

piecemeal out of it all like dandruff?
If he listens he can hear the faint scream of madness
echoing from subways,
fifty, forty-nine, forty-eight . . .

He can go no further.

OXFORD POETS

Fleur Adcock
Yehuda Amichai
James Berry
Edward Kamau Brathwaite
Joseph Brodsky
D. J. Enright
Roy Fisher
David Gascoyne
David Harsent
Anthony Hecht
Zbigniew Herbert
Thomas Kinsella
Brad Leithauser
Herbert Lomas
Derek Mahon
Medbh McGuckian
James Merrill
John Montague
Peter Porter
Craig Raine
Tom Rawling
Christopher Reid
Stephen Romer
Carole Satyamurti
Peter Scupham
Penelope Shuttle
Louis Simpson
Anne Stevenson
George Szirtes
Anthony Thwaite
Charles Tomlinson
Andrei Voznesensky
Chris Wallace-Crabbe
Hugo Williams

also

Basil Bunting
Keith Douglas
Edward Thomas